Lukas Scisly

Analytical Report

The relationship of factors that determine the price per square foot of a single family home in Wichita Falls, Texas

GRIN Verlag

Bibliografische Information der Deutschen Nationalbibliothek:

Die Deutsche Bibliothek verzeichnet diese Publikation in der Deutschen National-
bibliografie; detaillierte bibliografische Daten sind im Internet über http://dnb.d-
nb.de/ abrufbar.

Imprint:

Copyright © 2009 GRIN Verlag GmbH
Druck und Bindung: Books on Demand GmbH, Norderstedt Germany
ISBN: 978-3-640-76500-3

This book at GRIN:

http://www.grin.com/en/e-book/159427/analytical-report

GRIN - Your knowledge has value

Der GRIN Verlag publiziert seit 1998 wissenschaftliche Arbeiten von Studenten, Hochschullehrern und anderen Akademikern als eBook und gedrucktes Buch. Die Verlagswebsite www.grin.com ist die ideale Plattform zur Veröffentlichung von Hausarbeiten, Abschlussarbeiten, wissenschaftlichen Aufsätzen, Dissertationen und Fachbüchern.

Visit us on the internet:

http://www.grin.com/

http://www.facebook.com/grincom

http://www.twitter.com/grin_com

Midwestern State University

Dillard College of Business Administration

Wichita Falls, Texas

Analytical Report

The relationship of factors that determine the price per square foot of a single family home in Wichita Falls, Texas

Midterm Project

BUAD 5603 – Advanced Applied Business Statistics

Fall 2009

Submitted on October 20, 2009 by

<u>Lukas Scisly</u>

Table of Contents

List of Tables

1 Introduction

In Wichita Falls, Texas, a new real estate company was established. In order to become acquainted with the local residential market, the company requires a statistical analysis of the determinants that are likely to influence the price per square foot of single family homes in this area. For this purpose a consultant was commissioned. In order to investigate the information required by the real estate company, the relevant data was gathered from realtor.com, a real estate agent.

In order to conduct the analysis, at first the necessary data has to be collected. For this purpose a sample of fifty houses will be collected. The consultant decides to investigate nine variables that are likely to be factors in determining the asked price per square foot. Based on this data collection, a descriptive analysis will be conducted where the variables will be analyzed for their means, medians, standard deviations, as well as for their minimum and maximum values. The next step consists of conducting a correlation analysis where the relations between the variables will be investigated. Afterwards, a regression analysis will be conducted in order to find out whether the independent variables can explain the dependent variable. Finally, the findings will be summarized in a conclusion.

2   Statistical Analysis

2.1   Variable Determination

In order to figure out factors that might determine the asked price per square foot of a single family home in Wichita Falls, Texas, a detailed set of variables has to be established that can be measured and expressed in numbers. This set consists of a dependant variable and several independent variables that will be described in the following paragraphs.

2.1.1   Dependent Variable

A dependent variable is the observed result that is explained by one or several factors (i.e. independent variables). The dependant variable is the *price per square foot of a single family home.*

2.1.2   Independent Variables

The independent variables are possible determinants of the dependent variable (i.e. the price per square foot). These variables should not be influenced by each other. The following variables were chosen as potential determinants for the dependant variable:

-   *Age*: This factor was chosen because people might associate different levels of dilapidation with respect to the age of the house. Furthermore, newer houses are probably built using more modern architectural knowledge in terms of energy efficiency (e.g., heat insulation) and stability (e.g., hurricane secure). Those attributes could influence the house price per square foot.

- *Number of Bedrooms*: The quantity of bedrooms available in a single family home can be a major reason for a decision purchase a house. Especially, young families that plan to have many children might be more willing to pay higher prices for higher numbers of bedrooms in order to have enough space for their family.

- *Number of Bathrooms*: The quantity of bathrooms that a house possesses seems to be an important factor for a buying decision. Especially for families with several children, a higher number of baths might increase the quality of life in terms of convenience. Therefore, the number of bathrooms might affect the house price per square foot.

- *Living Room Size*: The living room is the place where the inhabitants, in many cases, spend the most of their time in order to communicate with each other and to round out the day. Furthermore, this room provides space to welcome guests and therefore, its size might be associated with wealth. Thus, the living room size might represent an influencing factor for the dependent variable.

- *Kitchen Size*: For several reasons, the kitchen size might influence the dependent variable. On the one hand, modern kitchens require more space than traditional kitchens. Furthermore, kitchens do not only provide a place to prepare food, but also have functions related to the living room. Often meetings with friends take place there. Therefore its size might represent a status symbol and thus influence the price per square foot.

2.1.3 Independent Dummy Variables

Dummy variables will be used to determine relationships between qualitative independent variables and the dependent variable. These variables can take a value of either one or zero, depending on whether a certain condition is available or not. The following variables were used as independent dummy variables:

- *Pool Availability*: A swimming pool can increase the value of a house because of the additional leisure activities it provides. Furthermore, it might represent a status symbol. Therefore, its availability might be related to the asked price per square foot. A value of one is assigned if a pool is available and zero otherwise.

- *Corner Lot*: A corner lot can provide advantages and disadvantages and therefore might influence the asked price per square foot. An advantage is that the house has two front sides to showcase, and therefore can make the house more prominent. Disadvantages can be, e.g., the substantially more sidewalk to clear of snow in winter or more noise caused by more traffic at crossroads. If the house's lot is a corner lot, this variable has the value of one; zero otherwise.

- *Family Room Availability*: A family room is an all-purpose room that can have similar functions as the living room does. However, its existence implies more value to the house because it provides more space for the family to perform more activities. Therefore, its availability might influence the price per square foot. This variable has the value of one if a family room exists and zero otherwise.

- *Quantity of Stories*: More than one story makes a house look more worthwhile but also can provide more obligations in terms of maintenance. Therefore, the quantity of stories can be an influencing factor of the dependent variable. If the house has more than one story, the variable has the value of one. If the house consists of only one story then this variable has the value of zero.

2.2 Descriptive Statistics

The following part of this paper will consist of an analysis of the variables introduced in part 2.1 with respect to their means, medians, standard deviations, as well as their minimum and maximum values. These measures base on the collected data available in the appendix (Table 5.1 and Table 5.2). Other measures than presented in this paragraph can also be seen in the appendix (Table 6).

2.2.1 Mean

By using the mean of a data set, it is possible to compare the value of a certain sample with the average of all samples used in that data set. With respect to the observations at hand, it can be demonstrated to what extent a certain variable of a certain observation varies within the same variable in all observations of the data set. In the following table, the means of all chosen determinants, as well as their meanings are illustrated:

Determinant	Mean	Meaning in terms of the sample
Price/ Square Foot	98.78	An average house costs $98.78 per square foot.
Age	38.66	An average house is 38.7 years old.
Bedrooms	3.68	An average house has 3.7 bedrooms.
Baths	2.99	An average house has 3 bathrooms.
Living Room in Square Foot	311.34	The average living room size is 311.34 square foot.
Kitchen in Square Foot	160.15	The average kitchen size is 160.15 square foot.
Pool	0.28	28 percent of all houses have a pool.
Corner Lot	0.24	24 percent of all houses are located at a corner lot.
Family Room	0.70	70 percent of all houses have a family room.
Additional Stories	0.40	40 percent of all houses have an additional story.

Table 1: Means and their meanings in terms of the sample.

2.2.2 Median

By using the median of a data set, it can be determined whether the value of a certain variable is smaller or larger than 50 percent of the values of the other observations. With respect to the dummy variables, it can be determinated whether more than 50 percent of the observations possess a particular attribute (e.g., a pool). In the following table, the medians of all chosen determinants, as well as their meanings are illustrated:

Determinant	Median	Meaning in term of the sample
Price/ Square Foot	95.00	50 percent of all houses cost $95.00, or more, per square foot; 50 percent cost less.
Age	39.00	50 percent of all houses are 39 years old or older; 50 percent are younger.
Bedrooms	4.00	50 percent of all houses have 4 or more bedrooms; 50 percent have less.
Baths	3.00	50 percent of all houses have 3 bathrooms; 50 percent have less.
Living Room in Square Foot	313.50	50 percent of all living rooms have a size of 313.5 square foot or are larger; 50 percent are smaller.
Kitchen in Square Foot	156.04	50 percent of all kitchens have a size of 156.04 square foot or are larger. 50 percent are smaller.
Pool	0.00	Less than 50 percent of the houses have a pool.
Corner Lot	0.00	Less than 50 percent of the houses have a corner lots.
Family Room	1.00	50 percent or more of the houses have a family room.
Additional Stories	0.00	Less than 50 percent of the houses have an additional story.

Table 2: Medians and their meanings in terms of the sample.

2.2.3   Standard Deviation

Using the standard deviation of a data set, it can be determined how far the most variables of the observation can be expected away from their mean. With respect to the dummy variables, the standard deviation does not need to be interpreted since the values of those variables only can have the values zero or one. In the following table, the standard deviations of all chosen determinants, as well as their meanings are illustrated:

Determinant	Standard Deviation	Meaning in terms of the sample
Price/ Square Foot	25.19	The most houses cost between $73.60 and $123.97 per square foot.
Age	23.99	The most houses are between 14.7 and 62.7 years old.
Bedrooms	0.74	The most houses have between 2.9 and 4.4 bedrooms.
Baths	0.87	The most houses have between 2.1 and 3.9 bathrooms.
Living Room in Square Foot	82.53	The most living rooms have a size between 228.8 and 393.9 square foot.
Kitchen in Square Foot	66.45	The most kitchens have a size between 93.7 and 226.6 square foot.
Pool	0.45	---
Corner Lot	0.43	---
Family Room	0.46	---
Additional Stories	0.49	---

Table 3: Standard deviations and their meanings in terms of the sample.

2.2.4   Minimum and Maximum

The minima and maxima of the variables show the smallest and the largest value respectively. In the following table, the minimum and the maximum values of all chosen determinants, as well as their meanings are illustrated. With respect to the dummy variables, the minima, as well as the maxima do not need to be interpreted since those values always have either the value of one or zero (i.e. an attribute does or does not exist).

Determinant	Minimum	Maximum	Meaning in terms of the sample
Price/ Square Foot	63.41	184.45	The smallest price per square foot is $63.41; the largest is $184.45.
Age	0.00	84.00	The youngest house is less than 1 year old; the oldest is 84 years old.
Bedrooms	2.00	5.00	The houses have at least 2 bathrooms; at the most 5.
Baths	2.00	5.00	The houses have at least 2 bedrooms; at the most 5.
Living Room in Square Foot	120.46	509.76	The smallest living room size is 120.46 square foot; the largest is 509.76 square foot.
Kitchen in Square Foot	19.42	353.08	The smallest kitchen size is 19.42 square foot; the largest is 353.08 square foot.
Pool	0.00	1.00	---
Corner Lot	0.00	1.00	---
Family Room	0.00	1.00	---
Additional Stories	0.00	1.00	---

Table 4: Minima and maxima and their meaninigs in terms of the sample.

8

2.3   Correlation Analysis

A correlation analysis can reveal relations amongst the dependent and the independent variables. However, the independent variables should be as unrelated as possible. To interpret the results, the correlation values between 0 and 0.4 were treated as small relations and values larger than 0.4 were interpreted as strong relations. The complete correlation matrix can be found in the appendix (Table 7). The following part explains the revealed strong correlations. The correlation coefficient is printed in parentheses:

- *Price per Square Foot*: In terms of the dependent variable, there is a strong relation with the number of bathrooms (0.5207) and with the kitchen size (0.5485). Both these independent variables seem to have a strong relation with the price per square foot. Whether these variables are significant will be investigated by conducting a regression analysis.

- *Number of Bathrooms*: A strong relation between this variable and the quantity of bedrooms (0.6154) could be revealed. Therefore, it is likely that an increasing amount of bathrooms causes a higher number of bedrooms, and vice versa.

- *Kitchen Size*: This variable is strongly related to the number of bathrooms (0.4461) and implicates that an increasing kitchen size is likely to cause a higher amount of bathrooms, and vice versa.

- *Additional Stories*: This variable has a strong relation with the quantity of bedrooms (0.5791) and with the number of bathrooms (0.5810). Thus, houses with additional stories are more likely to have more bedrooms and bedrooms, and vice versa.

2.4 Regression Analyses

In order to investigate whether the independent variables are meaningful determinants that help to explain the differences in the house prices per square foot, a regression analysis had to be conducted. The complete regression analysis can be found in the appendix (Table 8 and Table 9). In the following paragraphs the results of this analysis will be explained:

2.4.1 Regression Analysis I

Regarding the revealed strong correlation between the number of bathrooms and the number of bedrooms, the second variable (i.e. the number of bedrooms) was not included to the regression analysis.

With respect to the regression statistics, the *R Square Value* (0.4924) reveals that 49.24 percent of the changes in the price per square foot can be explained by changes in the independent variables. Due to the fact that the *R Square Value* can increase by using a large number of independent variables, the *Adjusted R Square Value* (0.3934) that considers the amount of observations gives a better result of how well the dependent variable is explained by the independent variables.

With respect to the revealed *Significance of F* (0.0002), only a 0.02% chance exists that the regression model fits only by accident. Thus, the regression model can explain significantly that changes in the dependent variable are caused by changes in the chosen independent variables.

In order to analyze which variables are significant, the *P-Value* of each variable has to be investigated. By applying a significance level of 5 percent, only the kitchen size in square foot (*P-Value*: 0.0021) can be derived as independent variable that significantly affects the dependent variable.

2.4.2   Regression Analysis II

Based on the information provided by the first regression analysis, a second regression analysis can be conducted, using only the dependent and the significant independent variable to conclude a linear function between these.

The second regression analysis reveals that the use of only one variable (i.e. kitchen size in square foot) is still significant; the *P-Value* of this variable is 0.0000. As the *Significance of F* decreased to 0.0000, it can be implied that the decrease in independent variables increased the level of significance in the model on hand. Furthermore, the smaller difference between *R Square* (0.3009) and *Adjusted R Square* (0.2863) reveals that the significance of the model is not based on the amount of independent variables.

Thus, the coefficients for the regression line are the intercept (65.4829) and the kitchen size in square foot (0.2079). In terms of the model, this means that a house without kitchen costs $65.48 per square foot. Every additional square foot in terms of the kitchen size causes an increase in the price per square foot by $0.21. By assigning the labels "Y" to the dependent variable (i.e. price per square foot of a single family home) and "X" to the independent variable (i.e. kitchen size in square foot), the regression formula is the following: Y = 65.4829 + 0.2079 * X.

3 Conclusion

In this statistical analysis fifty data sets were collected. Out of nine variables that were chosen to be possible determinants for the asked price per square foot of a single family home in the zip code area 76308 (Wichita Falls), eight were analyzed in terms of their significance. It could be revealed that only one of these variables (i.e. the kitchen size in square foot) is statistically significant for the estimation of the asked house price per square foot of a single family home.

Based on the findings, the price of a house per square foot that does not have a kitchen would be $65.48. The availability of a kitchen increases the price per square foot by $0.21 per square foot in terms of the kitchen size.

Although the investigated relation is highly significant, factors other than the analyzed variables could also have an important influence on the price per square foot of a single family home. Such factors could be the distance to favored grocery stores, churches, schools, as well as the distance to the airport (with respect to the noise). Furthermore, the individual perception of the neighborhood could be a factor, too. However, the derived regression formula can provide good values that can be used by the real estate company in order to become acquainted with the local residential market.

4 References

www.realtors.com.

John E. Hanke, Dean W. Wichern (2009), *Business Forecasting*, ninth edition.

5 Appendix

5.1 Raw Data

Name	Price	Square Feet	Price/ Square Feet	Year Built	Age	Stories	Bed-rooms	Baths	Living Room in Square Foot	Kitchen in Squa-re Foot	Pool	Corner Lot	Family Room	Office	Additional Stories
#5 White Oak C.	199,900	2,577	77.57	1985	24	1	4	2.5	368.57	132.75	0	0	1	0	0
#6 Jeffus C.	174,500	2,017	86.51	1993	16	1	3	2	238.00	160.00	0	0	1	0	0
#9 Breezewood C.	359,900	3,393	106.07	2003	6	1.5	4	3.5	351.31	207.83	0	0	1	0	1
2 Olive Branch C.	379,900	3,061	124.11	2009	0	2	4	3	302.50	173.25	0	0	1	0	1
2000 Granada D.	149,900	1,880	79.73	1980	29	1	3	2.5	340.75	108.00	1	0	0	0	0
2006 Avondale S.	575,000	4,319	133.13	1929	80	2	4	3.5	391.00	162.00	1	0	1	0	1
2010 Irving S.	435,000	4,520	96.24	1957	52	2	4	4.5	285.00	122.13	1	0	1	1	1
2018 Berkeley D.	345,000	3,264	105.70	1935	74	2	4	3.5	309.83	213.75	0	0	1	0	1
2025 Clarinda A.	1,179,000	6,392	184.45	1929	80	2	3	3.5	406.50	353.08	1	1	1	0	1
2029 Peachtree L.	330,000	2,952	111.79	1993	16	2	4	3	276.94	120.75	1	1	1	0	1
2105 Miramar S.	595,000	5,508	108.02	1962	47	2	4	3.5	231.83	213.75	0	0	1	1	1
2109 Ellingham D.	875,000	5,036	173.75	1939	70	2	4	3.5	386.75	168.33	0	1	1	0	1
2200 Miramar S.	475,000	5,579	85.14	1951	58	2	4	4	377.36	221.51	0	1	1	0	1
2206 Midwestern Pw.	410,550	3,570	115.00	1964	45	1	3	3.5	274.42	152.08	1	0	1	0	0
2300 Kirk D.	355,000	3,300	107.58	1958	51	1	3	3	351.00	195.56	0	1	1	0	0
2304 Ellingham D.	329,900	3,858	85.51	1951	58	1	3	3	317.17	182.50	0	0	1	0	0
2309 Miramar S.	399,500	4,091	97.65	1928	81	2	4	3	480.00	115.94	0	0	1	0	1
2310 Brook Hollow D.	249,300	2,870	86.86	1978	31	1	4	3	337.79	89.38	1	0	1	0	0
2310 Ellingham D.	375,000	3,491	107.42	1949	60	2	5	4.5	358.36	192.00	0	0	1	1	1
2310 Irving Pl.	329,000	3,111	105.75	1960	49	1	3	2.5	296.08	200.74	0	0	1	1	0
2402 Bryan Glen S.	182,000	2,366	76.92	1981	28	1	3	2.5	509.76	96.09	0	0	0	0	0
2406 Buena Vista W.	325,000	4,720	68.86	1937	72	2	5	3.5	393.13	170.50	0	1	1	0	1
2407 Merrimac D.	199,900	2,837	70.46	1969	40	1	4	2.5	214.42	149.89	1	0	1	0	0
2415 Ellingham D.	179,500	2,222	80.78	1940	69	1	3	2	270.00	123.75	0	0	1	0	0
2500 N Elmwood Ci.	370,000	4,250	87.06	1983	26	2	5	3.5	281.17	270.52	0	0	1	1	1

Table 5.1: Raw Data Part I.

14

Name	Price	Square Feet	Price/ Square Feet	Year Built	Age	Stories	Bed- rooms	Baths	Living Room in Square Foot	Kitchen in Squa- re Foot	Pool	Corner Lot	Family Room	Office	Additional Stories
2504 Beefeater D.	146,500	1,874	78.18	1961	48	1	2	2	406.67	120.00	0	1	0	0	0
2620 Amherst S.	165,000	2,123	77.72	1957	52	1	3	2	262.50	189.00	0	0	1	0	0
2628 Amherst D.	209,900	2,448	85.74	1975	34	1	4	3	135.06	73.51	0	1	1	0	0
2722 Chase D.	163,000	1,872	87.07	1971	38	1	3	2	361.96	117.02	0	0	0	0	0
2805 Elmwood A.	144,500	1,861	77.65	1972	37	1	3	2	274.94	100.31	0	0	0	0	0
2805 Happy Hollow D.	157,500	2,187	72.02	1979	30	1	3	2	360.31	109.25	0	0	0	0	0
2810 S Shepherds Glen	189,900	2,039	93.13	2001	8	2	4	2.5	120.46	91.04	0	0	1	0	1
2818 Happy Hollow D.	141,500	1,696	83.43	1981	28	1	2	2	273.24	130.00	0	0	1	0	0
2937 Loma Linda L.	199,900	2,031	98.42	2008	1	1	3	2.5	392.71	132.00	0	0	0	0	0
3002 Hamilton B.	998,500	10,650	93.76	1925	84	2	5	4.5	241.44	229.33	0	1	1	1	1
3207 Taft S.	184,500	2,279	80.96	1955	54	1	3	2	342.36	29.25	0	1	1	0	0
3209 Martin B.	369,000	3,806	96.95	1952	57	1	3	4	360.00	170.00	0	0	1	1	1
3300 Harrison S.	675,000	5,079	132.90	1966	43	1	4	5	322.50	242.67	1	0	1	0	0
3304 Robin L.	425,000	3,970	107.05	1966	43	1	3	3	176.76	222.30	0	0	1	0	0
3528 Cranbrook L.	135,000	1,882	71.73	1965	44	1	4	2	230.22	104.25	0	0	0	1	1
4111 Seabury D.	134,000	2,028	66.07	1980	29	2	3	2	228.08	113.33	0	0	0	1	1
4155 Candlewood Ci.	259,000	2,490	104.02	2007	2	1	4	2.5	292.50	187.50	0	0	0	0	0
4207 Driftwood	725,000	5,411	133.99	1996	13	2	4	4.5	221.67	342.83	1	0	1	1	1
4500 Meadowbrook D.	139,000	2,192	63.41	1957	52	1	4	2	326.24	133.07	0	1	0	0	0
4806 Bridge Creek D.	419,900	3,478	120.73	2003	6	2	5	3.5	189.00	216.00	0	0	1	0	1
4810 Bridge Creek D.	318,250	2,573	123.69	2004	5	1	4	3	351.00	207.67	0	0	0	0	0
4812 Maplewood A.	212,900	2,594	82.07	1972	37	2	4	2.5	217.78	132.22	0	0	1	1	0
5 Joy Court	605,000	5,300	114.15	2004	5	2	5	5	380.00	19.42	0	0	0	0	0
8 Breezewood C.	370,000	2,758	134.16	2002	7	1	4	3	465.15	212.33	0	0	0	1	1
806 Royal R.	173,500	1,770	98.02	1995	14	1	3	2	284.67	87.00	0	0	0	0	0

Table 5.2: Raw Data Part II.

15

5.2 Descriptive Analysis

	Price/ Square Foot	Age	Bed-rooms	Baths	Living Room in Square Foot	Kitchen in Square Foot	Pool	Corner Lot	Family Room	Additional Stories
Mean	98.78	38.66	3.68	2.99	311.34	160.15	0.28	0.24	0.70	0.40
Standard Error	3.56	3.39	0.10	0.12	11.67	9.40	0.06	0.06	0.07	0.07
Median	95.00	39.00	4.00	3.00	313.50	156.04	0.00	0.00	1.00	0.00
Mode	#N/A	52.00	4.00	2.00	351.00	213.75	0.00	0.00	1.00	0.00
Standard Deviation	25.19	23.99	0.74	0.87	82.53	66.45	0.45	0.43	0.46	0.49
Sample Variance	634.39	575.62	0.55	0.75	6,811.69	4,415.15	0.21	0.19	0.21	0.24
Kurtosis	2.50	-0.90	-0.26	-0.35	0.09	1.18	-1.02	-0.44	-1.24	-1.90
Skewness	1.35	0.10	-0.03	0.65	-0.02	0.63	1.01	1.26	-0.90	0.42
Range	121.04	84.00	3.00	3.00	389.30	333.67	1.00	1.00	1.00	1.00
Minimum	63.41	0.00	2.00	2.00	120.46	19.42	0.00	0.00	0.00	0.00
Maximum	184.45	84.00	5.00	5.00	509.76	353.08	1.00	1.00	1.00	1.00
Sum	4,939.11	1,933.00	184.00	149.50	15,566.82	8,007.39	14.00	12.00	35.00	20.00
Count	50.00	50.00	50.00	50.00	50.00	50.00	50.00	50.00	50.00	50.00

Table 6: Descriptive Analysis.

5.3 Correlation Analysis

	Price/ Square Foot	Age	Bed- rooms	Baths	Living Room in Square Foot	Kitchen in Square Foot	Pool	Corner Lot	Family Room	Additional Stories
Price/ Square Foot	1.0000									
Age	0.0328	1.0000								
Bedrooms	0.1883	-0.0200	1.0000							
Baths	0.5207	0.1737	0.6154	1.0000						
Living Room in Square Foot	0.1825	0.1988	-0.1357	0.0941	1.0000					
Kitchen in Square Foot	0.5485	0.1562	0.2244	0.4461	-0.0245	1.0000				
Pool	0.2532	0.0502	-0.0316	0.2931	0.0070	0.0495	1.0000			
Corner Lot	0.1691	0.3708	0.1175	0.1704	0.1061	0.0945	-0.0375	1.0000		
Family Room	0.2389	0.3875	0.2500	0.3996	-0.2857	0.3682	0.1166	0.1635	1.0000	
Additional Stories	0.3857	0.1612	0.5791	0.5810	-0.0072	0.3192	0.0364	0.1147	0.3563	1.0000

Table 7: Correlation Analysis.

5.4 Regression Analysis I

Regression Statistics

Multiple R	0.7017
R Square	0.4924
Adjusted R Square	0.3934
Standard Error	19.6176
Observations	50

ANOVA

	df	SS	MS	F	Significance F
Regression	8	15306.1137	1913.2642	4.9715	0.0002
Residual	41	15778.8384	384.8497		
Total	49	31084.9522			

	Coefficients	Standard Error	t Stat	P-value	lower 95%	upper 95%	lower 90%	upper 90%
Intercept	36.2368	15.5452	2.3311	0.0247	4.8426	67.6310	10.0761	62.3975
Age	-0.2295	0.1429	-1.6057	0.1160	-0.5180	0.0591	-0.4699	0.0110
Baths	4.4769	4.6839	0.9558	0.3448	-4.9824	13.9362	-3.4055	12.3593
Living Room in Square Foot	0.0696	0.0391	1.7830	0.0820	-0.0092	0.1485	0.0039	0.1354
Kitchen in Square Foot	0.1601	0.0487	3.2870	0.0021	0.0617	0.2584	0.0781	0.2420
Pool	10.4226	6.6279	1.5725	0.1235	-2.9627	23.8079	-0.7313	21.5765
Corner Lot	8.0283	7.0824	1.1336	0.2636	-6.2749	22.3314	-3.8905	19.9470
Family Room	4.0372	8.1397	0.4960	0.6225	-12.4012	20.4757	-9.6609	17.7353
Additional Stories	7.5997	7.1756	1.0591	0.2958	-6.8917	22.0911	-4.4759	19.6754

Table 8: Regression Analysis I.

5.5 Regression Analysis II

Regression Statistics

Regression Statistics	
Multiple R	0.5485
R Square	0.3009
Adjusted R Square	0.2863
Standard Error	21.2777
Observations	50

ANOVA

	df	SS	MS	F	Significance F
Regression	1	9353.3638	9353.3638	20.6594	0.000037
Residual	48	21731.5884	452.7414		
Total	49	31084.9522			

	Coefficients	Standard Error	t Stat	P-value	lower 95%	upper 95%	lower 90%	upper 90%
Intercept	65.4829	7.9200	8.2680	0.00000000	49.5586	81.4073	52.1993	78.7666
Kitchen in Square Foot	0.2079	0.0457	4.5453	0.00003720	0.1159	0.2999	0.1312	0.2847

Table 9: Regression Analysis II.

19